The Forgotten Scientist

First published by Jacana Media (Pty) Ltd in 2020
Second impression 2021

10 Orange Street
Sunnyside
Auckland Park 2092
South Africa
+2711 628 3200
www.jacana.co.za

© Lorato Trok, 2020

All rights reserved.

ISBN 978-1-4314-2989-9

Layout by Aimee Armstrong
Editing by Nkhensani Manabe
Proofreading by Lara Jacob
Set in Goudy Oldstyle Std 12/14pt
Job no. 003672
Printed and bound by Print on Demand

See a complete list of Jacana titles at www.jacana.co.za

The Forgotten Scientist
The Story of Saul Sithole

Lorato Trok

In memory of Saul Sithole, the black scientist who was never recognised as a scientist.

Contents

Introduction	1
1 Family life	5
2 Dedicated to service	8
3 Historical findings	15
4 Retirement	23
5 A life well lived	26
6 Flying with the birds	30
Afterword	32
About the author	38
Acknowledgements	39
Glossary	40
Sources	43

Introduction

Ornithology is the study of birds. Ornithologists are scientists who devote their time to understanding all aspects of bird life, including their identifications, bird calls, anatomy and physical appearance, as well as flight patterns.

Ornithology on the African continent is as ancient as Africa itself. Africans used their traditional and indigenous knowledge and perspectives in collecting and recording ornithological perspectives. With the scramble for Africa by Europeans, ornithology on the continent moved away from indigenous knowledge to white ornithologists named as collectors, even if they lifted no finger in shooting and preparing the birds.

Europeans used scientific theories to drive ideas of racial superiority when they colonised African states. African men were only ever used as assistants to white ornithologists, even though they had long been a part of birding culture in their respective countries. These assistants were rarely mentioned in the official state archival record.

Yokana Kiwanuka of Kenya does not appear in published narrative accounts even though a sunbird was named after him, *Anthreptes reichenowi yokonae*. Very little is known about Kiwanuka's investment and participation in the network of birders.

One of the very best assistants to ornithologists Jack Vincent and Admiral Hubert Lynes is mentioned in several publications as "Ali". Vincent expressed affection and personal admiration but at the same time exhibited obsession with racial status. He never recorded Ali's story or surname. It was only later, from other sources that it was known that Ali's surname was Safi, and that he became a migrant from Malawi. Ali Safi was not just an assistant as his managers perceived, but one of Africa's finest birders. He worked at the zoo in Pretoria and in 1930 he joined the Vernay-Lang Kalahari Expedition with Lang as scientific leader. In 1931 he was recommended to Lynes and Vincent as an assistant. With them he travelled to present-day Zimbabwe, Zambia, Malawi, Angola, Mozambique and the Democratic Republic of Congo. His story was only entered into the archival records in 1933.

South Africa's birding space was completely dominated by Europeans. Saul Sithole, like many African men of his generation, was defeated by colonialism and apartheid. Information on birding in Africa is solely ascribed to white men, regardless of how much the likes of Saul Sithole, Matthews Mathabathe, Ali Safi and others contributed to this unfamiliar and important scientific knowledge. They were still just "useful natives" in a noble profession reserved for white honour and prestige.

Sithole spent more than 60 years of his life at the Transvaal

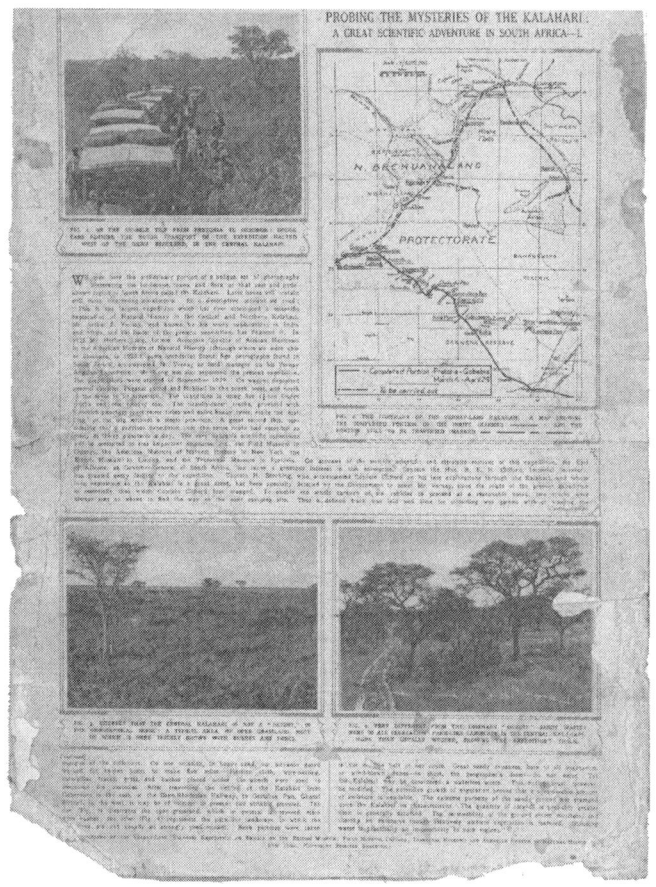

A 1930 newspaper article with a map of the expedition across Southern Africa

Museum (now called the Ditsong National Museum of Natural History), more than half of his life as a birder and in various capacities, yet no history book nor books about ornithology in South Africa mention what his skills contributed to ornithology and anthropology.

As the world changes and more people are embracing their history with no political restrictions, more and more stories of our society's hidden figures are unearthed and are being given space in the mainstream. Saul Sithole is one of the hidden figures whose story of scientific knowledge in ornithology deserves praise and honour. This book tries to honour his legacy and celebrate his decades-long career in ornithological work in Africa. It is also a love letter to the many men and women in South Africa and on the African continent whose stories of courage were muffled by decades of injustice, exclusion and dispossession.

This book brings to light the contribution of Saul Sithole to the scientific world of ornithology, palaeontology and anthropology from the 1920s up until his retirement in 1990.

1
FAMILY LIFE

Saul Sithole was born on 20 September 1908 in Standerton, Mpumalanga. The Sithole family was Zulu speaking, although it was believed that they were Tsonga. His father was a Methodist lay preacher.

Saul was the second of three boys, between Solomon and Amos. He completed standard six (Grade 8), which was a noteworthy achievement for black South African children and would have qualified him as a teacher. He started his schooling at Khuzwayo Nkomo, one of the notable Methodist schools in Lady Selborne, a freehold black township in Pretoria.

Growing up, trapping birds was a way of life for boys in the township. Every single day Saul, his brothers and their friends would play this game. They would use the lid of a trash can, lift it up with a stick and put some breadcrumbs underneath it. A long string would be attached to the lid and the boys would hide behind a tree or a nearby hideout.

Saul Sithole and his mother

As soon as birds got in to get to the breadcrumbs, they would pull the string and trap the birds. Some days they would use slingshots to kill the birds. The boys did this mainly as a form of entertainment and as a way to catch birds for meat, but years later birds would define Saul Sithole's life and legacy in a different and more fulfilling way.

Saul's education was interrupted when his father died. He and Solomon went to work to enable young Amos to continue his studies. He worked periodically as a bus conductor until a friend of the family, Elephas Lebelo, helped Solomon and Saul find jobs at the Transvaal Museum.

Sithole's wife, Sophia Nomvula, was also from Lady Selborne and like him, had also studied till standard six.

The couple lived in Lady Selborne, where their only child Zondi was born in 1931. The Sitholes were forcefully removed to the township of Mamelodi, east of Pretoria, when the apartheid government intensified its laws of forced removals to resettle whites.

Saul and his wife Sophia settled well into their family life in Mamelodi where they brought up their daughter Zondi. Like other townships in South Africa, Mamelodi was on the outskirts of town, which was the primary purpose of white minority rule. Black people made daily commutes into town to get to work. Saul Sithole was no different. He commuted more than 60 kilometres daily to get to his work at the museum and back home.

Little Zondi started her primary schooling at the same school her father began his education, at Khuzwayo Nkomo in Lady Selborne. They later enrolled her at a Methodist boarding school in Hebron; she went on to complete her teaching diploma at Hebron College of Education. Zondi became a teacher and remained so for more than 30 years before her retirement at the age of 60 in 1991.

2
DEDICATED TO SERVICE

Saul Sithole started working at the museum in Pretoria on 11 November 1928. It is said that he started out as a cleaner. Saul proudly recalled helping to mount the elephant skeleton that has long been displayed at the entrance of the museum during his first year there.

Sithole's professional life became specialised around birds when he joined the Vernay-Lang Kalahari Expedition of 1930, a cooperative effort between the American Museum of Natural History in New York and the Transvaal Museum.

Saul Sithole was an ornithological assistant to Herbert Lang, a German-born mammologist, naturalist and photographer, who had worked at the American Museum of Natural History and is best known as the leader of the 1909 to 1915 expedition to the Congo rainforest. During Lang's next expedition to Angola, he remained in Africa and took up a position at the Transvaal Museum in 1927.

Saul Sithole (left) with his team and their bird specimen, Vernay-Lang Kalahari Expedition

Saul Sithole learned to skin birds from Herbert Lang. Even though there were other black ornithological assistants on the team, lead ornithologist Austin Roberts singled out Sithole's excellent work when he wrote in his report:

> "Lang is training the boys to make good skins ... Saul is far better than the other skinners and is doing very fine skins."

Even though the skinners, including Saul Sithole, were doing exceptional work, to their masters and ornithologist they were still described as "natives" and "boys". Their value was described in terms of their discipline and usefulness, not much as scientists who rendered valuable scientific work. This is how Lang described Saul Sithole's leadership:

> "This native, by his ready willingness to render himself useful and by his good example, assisted in maintaining an excellent discipline and great industry among the skinners."

Saul Sithole was a city slicker who struggled to establish himself in rural craft during fieldwork. He was an expert in using fine steel, cotton wool, alcohol and arsenic. He would read vigorously to learn about the trade of skinning and preserving birds shot in the field. Sithole would learn the same techniques his superiors learned, honing his technical skills on the job and gaining his bird skinning skills on expeditions where he could observe living birds.

His work with specimens was equal to no one and very precise. His daughter Zondi Zitha recalled her father telling her that once when one of his colleagues did a taxidermy very quickly, Sithole was called in to redo it because it was evident that it did not have his touch. One of Saul Sithole's co-workers, Matthews Mathabathe, who joined the museum in 1958 explained Sithole's work ethic better when he said, "He had artistic hands on his work. He was a handyman who could work with any scientist." One of Sithole's closest colleagues, Tamar Cassidy, explained that the skill of skinning and preparing birds required a feel for what the bird should look like; Saul Sithole had that sense. Cassidy expanded on this in a conversation with Professor Nancy Jacobs:

> "It's more than a skill. Every bird is different. You have to know the bird. There is certain craft involved in knowing how to deal with that kind of bird or skeleton or wing or whatever it is. The whole point is to make it

Saul Sithole (right) in Zululand, 1932

look presentable, and that is the hard part. Every bird is differently shaped. You have to know the muscles. You have to know the structure of the body. So that when you put it together again, it has to fall into position."

Sithole was the only assistant who mastered this art. After his expedition with Vernay and Lang, Sithole worked primarily with South Africa's best-known ornithologist

Saul Sithole (middle) with Austin Roberts (right) and a local man in Zululand during the 1932 Zululand expedition with a hippo they trapped

Saul Sithole (right), Samson Maseko (left) and an unidentified expedition leader

Austin Roberts as his general assistant and chief companion. Austin Roberts is synonymous with birding in South Africa and advocating for indigenous understandings of bird behaviour as a useful tool for collectors. His *Birds of South Africa*, published in 1940, was the most successful book in ornithology and features names of birds in African languages. Roberts was not without controversy. He intimated that Africans (whom he called "observant natives") understood birding through superstition and nothing else.

Sithole accompanied Roberts on collecting trips to Vryburg, Grahamstown and a six-week expedition to Zululand (KwaZulu-Natal) in 1932. Roberts was so enamoured to Sithole that he wrote glowingly about his work in all his reports, one of which was a February–March 1933 report where he wrote:

> "Native Assistant Saul has continued his useful assistance in cleaning skulls and re-making up badly prepared skins for birds."

Saul Sithole became a valuable companion to Roberts on the 1932 expedition to Zululand. Roberts did not speak Zulu, nor did he understand the local culture. Zondi Zitha recalled that there were times when the locals would not want to help Roberts. She remembered her father telling her about a hippo they trapped in Zululand that the locals would not help to get out of the trap. He went out and pleaded with them in Zulu and they acquiesced. Of course, Sithole was not only helping with the language, he was involved in the science of skinning the birds.

During the 1930 Vernay-Lang expedition, Lang described

Sithole as a leader. He was able to lead the team of skinners to diligently work with precision and dedication to produce as many bird skins as possible. This expedition was one of the best as they produced, among other things, thirty thousand bird skins for four museums, namely the Transvaal Museum, the Field Museum of Chicago, the American Museum of Natural History in New York, and the British Museum of Natural History (Natural History Museum) in London.

3
HISTORICAL FINDINGS

Sithole learned many skills and he was given new responsibilities at the museum working with fossils as his work was so precise. Until 1934 he had spent six years on mammals and birds, preserving and preparing them for displays. Sithole's skills were incomparable. He even took on the responsibility of transferring his skills to high-level professional employees of the museum.

Professor Francis Thackeray, the anthropologist who would later become the director of the Transvaal Museum, learned to prepare specimens from Saul Sithole in 1971, when he was almost 20 years old. Professor Thackeray recalled that he and Sithole used to work together at the back of the museum. For obvious reasons, one laboratory had the undeniable smell of dead animals in a large tank filled with water. Sithole made a concoction that he said would offset the stench of the rotting meat. Thackeray remembers going into the room the next day and to his surprise, the smell was gone: "Saul would always say, 'just one drop, Francis', and it would work".

Animal specimens prepared by Saul Sithole featured in a 1930 newspaper article

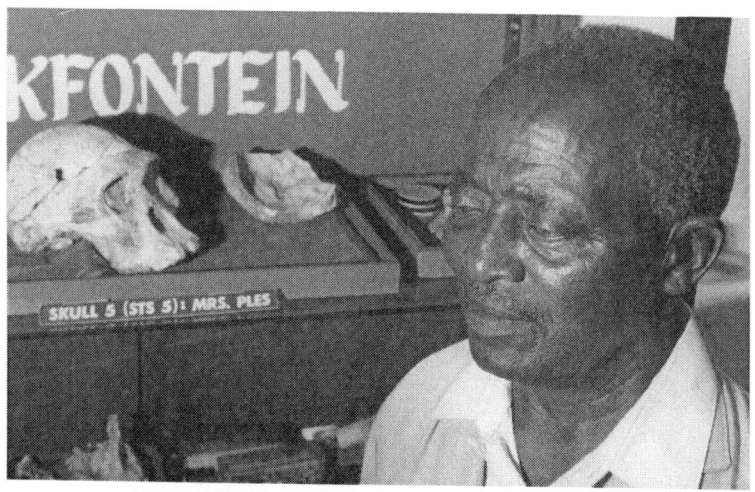

Saul Sithole with a replica skull of Australopithecus africanus *known as "Mrs Ples" at Ditsong National Museum of Natural History*

Sithole was involved with many palaeo-anthropological findings while accompanying his supervisors as an assistant. He worked with Dr Robert Broom, a medical doctor who excavated at the Sterkfontein limestone caves. He was present with Broom in August 1936 when the second ever *Australopithecus africanus* skull (TM 1511) was discovered. It was known as "Mrs Ples" (the first to be found was the "Taung Child").

Zondi Zitha recalled that her father was very proud of his zoological and paleontological work at Sterkfontein. He became invested in the challenge, she recalled.

During the 1937 Barlow-Transvaal Museum expedition to South-West Africa (Namibia) with Roberts and the herpetologist Vivian FitzSimons, Sithole collected insects and trapped small mammals and birds. He was pleased to

Saul Sithole (far right), Vivian FitzSimons (middle) and Austin Roberts (left), with prepared bird specimens collected in South-West Africa, 1937

have trapped two of the rare birds while Austin Roberts only trapped one. His proudest moment, which he will later relate to entomologist Charles Koch during their trip to Angola in 1956, was when he showed Austin Roberts how to collect black and white striped beetles in the dunes of Namibia. Roberts acknowledged Sithole's rare find in the South African ornithological journal called the *Ostrich*.

In 1938 he travelled with Dutch scientists to the Kruger National Park to collect blood samples from giraffes. From 1939 to 1940 Sithole and Austin Roberts took a four-month expedition trip to the southern Cape where they were involved in a car crash that knocked him unconscious and injured Roberts slightly. He travelled very little after this trip. He would concentrate on improving the quality

The expedition team often had to cross dangerous territories (Ngonye Falls, Western Zambia, July 1952)

of the specimens prepared by others instead. After the successful 1937 expedition with Austin Roberts and Vivian FitzSimons in South West Africa, Saul Sithole undertook a few more expeditions across the African continent, and returned back to South West Africa during 1952 zoological expedition to Katima Mulilo.

Saul made his last two expeditions for the museum in 1956. His trusted confidante Austin Roberts had died from a heart attack while driving in 1948, while Robert Broom had ceased to work at the Sterkfontein caves. Saul's last expedition was to the Blouberg region of what is now Limpopo with ornithologist O.P.M. Prozesky, where he squeezed in a visit to the Rain Queen. Saul was gifted a sheep to slaughter.

Convoy between Katima Mulilo (Namibia) and Ngonye Falls (Zambia), July 1952

Nangweshi, Western Zambia, July 1952

Science and racial segregation

As skilled as Saul Sithole was in scientific work, even going as far as teaching his white senior colleagues who were seasoned scientists some tools of the trade at the museum and on expeditions, he could not become a scientist because he was black.

Sithole was different from his black colleagues who worked at the museum. Makawe's and Mathabathe's understanding of birds was in indigenous knowledge, whereas Sithole worked in a more scientific capacity with birds and fossils. He was close to white scientists and thus the knowledge of science. He still had to know his place as a black man in a segregated South Africa. His many expeditions with white scientists only offered him relief from carrying a Dompas, but he still had to sleep under a tree with a knife during these expeditions, unlike the scientists who carried government-sponsored guns for protection and slept under the safety of tents.

Zondi Zitha believed that had her father had the opportunity, he would have enjoyed a deeper engagement with science. Even Zitha herself could have followed in her father's footsteps and become a scientist. She was well-versed and had an interest in her father's work and knew as much about ornithology and anthropology from him. He pointed out rock formations and bird calls to his daughter. He created his own cabinet of natural history at his home, with a replica of "Mrs Ples".

There is no record of Saul Sithole challenging the authority of the white scientists he travelled with on matters of race and segregation. The only record of him being angry at authority was when he was blamed for the car accident that injured Roberts and a subsequent ban on his driving.

Far left, George Barlow, mine manager at Sterkfontein, behind him is Saul Sithole, who was Dr Broom's assistant during the 1936 excavations at Sterkfontein. Front right, Dr Robert Broom, palaeontologist, pointing to the position where part of skull of Australopithecus *(TM 1511) was discovered in 1936*

4
Retirement

Saul Sithole gave 62 years of his life working for the Transvaal Museum. His incomparable skills as an ornithological assistant and working on fossils require a place in the history books.

South Africa's history of racial oppression to the majority of its population through colonial rule and repressive apartheid laws dashed the dreams of many African men and women whose contributions are forgotten or credited to the white people they worked with. Race was an impediment to what Sithole could have been. Despite his contribution to the scientific knowledge and all his accomplishments, he was still a black man in South Africa, and in a white man's territory of science.

The pay disparities between blacks and whites were enormous. Although Sithole enjoyed working at the museum compared with the other options for black men at the time, such as working in deadly mines, his salary was not better off than that of the miners. In a 1935 proposal for a vertebrate animal survey, Austin Roberts budgeted £800 per year for the administrative officer, £500 per year for the senior field officers, jobs that were reserved for whites, and only £45 per year for "native" assistants, the same salary as an unskilled black labourer at the mines.

Museum correspondence in 1956 lists the highest pay level for a preparator at £180 per year, double what an unskilled black railway worker or migrant gold miner could earn, which was still not half what an unskilled white railway worker could earn. Two shillings from these paltry salaries went towards Dompass fees every month. Museum workers also did not have pension. This frustrated Sithole and he decided to leave the museum to start his own business at the age of 50, in 1958.

Lady Selborne at the time had an entrepreneurial boom amongst the community and Sithole decided to tap into it. He thought that he could make more money working for himself than earning the paltry salary with no pension for these many years he had already been there. He started his own coal business that lasted for ten years before he and his family were forcibly removed to Mamelodi. His business did not survive the upheaval and he turned to the museum.

In 1968, Dr Robert Brain invited Sithole to return to the museum as a paid staff member.

Sithole was earning R534 a year in 1970, the lowest salary he had ever received. He was grateful that museum work did not require of him what other jobs reserved for black men demanded – physical labour. He was responsible for the display and taxidermy at the museum. He was not as strong as he used to be when he was younger.

He enjoyed being a preparator all those years back and he took on this task again when he returned to the museum. He worked in the outside room of the museum where he preserved new specimens. He wasn't as productive as he used to be, but he was useful to the museum as a guide. He was the first person tourists asked for when they came to visit the museum and he was happy to talk them through museum work and about his own accomplishments.

5
A LIFE WELL LIVED

Throughout his life, he remained active in the Methodist church and was a member of *Amadodana Ase Wesile*. He was a respectable man in his community.

His younger colleagues recall that he started a savings stokvel at the museum and preached frugality and a saving culture to them. Because he didn't have a pension, he saved for his own retirement. He used his own savings to erect tombstones for his parents, an act regarded as noble in the black community. He emphasised the importance of education to his grandchildren and great-grandchildren as well as his younger colleagues at the museum.

He was always professionally dressed and always carried a briefcase. Respectability was important to him, but it was also a virtue instilled in black men by Europeans to turn them docile in order to not mobilise against white rule. Saul Sithole respected hierarchy and did his work in silence. He did not believe in organised unions.

His eldest granddaughter Divine remembers her grandfather as someone who took care of his health and well-being. He encouraged her and her younger sister Wally to be active and take care of their health. Divine lived on the west side of Mamelodi whereas her grandparents lived on the east side, but each day without fail Saul Sithole would take that daily walk of more than 20 kilometres to visit her, even in his old age.

Sithole used his closeness to white colleagues for his own advantage when it mattered the most. With the help of one of his white managers at the museum, he enrolled Zondi at Hebron College of Education. Elite missionary schools that gave black children quality education were closed down by the National Party when it came to power, and they enforced a system of Bantu Education to black learners, an inferior education to prepare them for a life of servitude to the white people.

Divine said that her childhood memories are filled with images of her grandfather working on bird skins at home. Themba Zitha, the first great-grandchild of the Sitholes, Divine's son, was his great-grandfather's trusted confidante and was named after Saul. Themba remembers his great-grandfather carrying him everywhere when he was younger. He has kept all the specimens his great-grandfather collected on his travels and objects discarded by the museum as family heirlooms.

Themba was also Zondi Zitha's eldest grandchild and they were as tight as clay. He took the role of a family patriarch in the absence of his grandfather Napoleon and his great-grandfather Saul. Zondi and Napoleon were married in 1951 and had two daughters. They lived in Mamelodi before

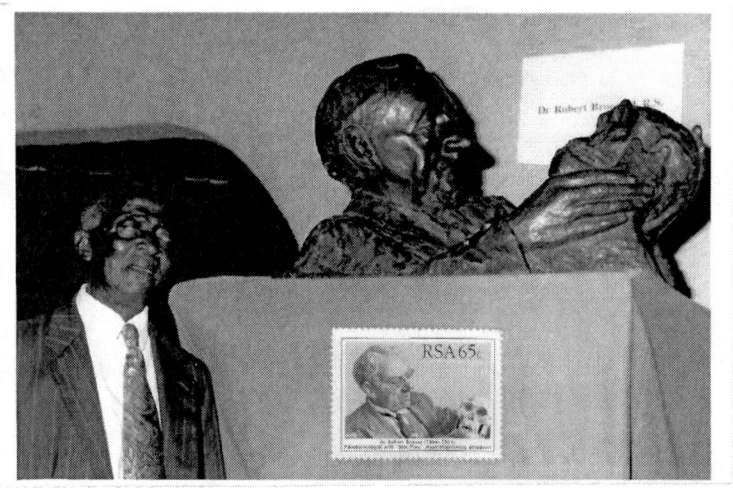

Saul Sithole with a bust of Dr Robert Broom who is shown holding the reconstructed skull of Australopithecus *("Plesianthropus") that had been discovered by Dr Broom with the assistance of Saul in 1936 at Sterkfontein. This photograph was taken by Dr C.K. (Bob) Brain when a stamp was launched to commemorate the palaeontological discoveries that had been made by Dr Broom.*

Napoleon died in 2008 at the age of 90, after 57 years of marriage. He (Napoleon), like everyone who crossed paths with Saul Sithole, was invested in his father-in-law's work at the museum and soaked up the knowledge from him.

Saul mulled over retirement at the age of 80 but Brain talked him into staying at the museum until they both retired. He eventually retired in December 1990. Sithole's family claims that after he gave 62 years of his life to the museum, diligently without fail, he was only gifted a wristwatch as a token of appreciation for his work.

He lived his retirement years doting on his grandchildren and great-grandchildren, never failing to tell them stories about his decades-long service at the museum.

6
FLYING WITH THE BIRDS

The museum named its staff service award after Saul Sithole in 1997, a deserved accolade to restore his honour and acknowledge his contribution in the fields of ornithology and archaeology. The award recognises staff members who have a long-service record with the museum. Having worked more than six decades at the museum, Sithole's grandchildren and great-grandchildren could not escape his professional life at home. What Themba Zitha remembers about his great-grandfather, besides the love he showered him and his whole family with, is how his love for birds and museum work came close to eclipsing anything else he held dear to his heart. He talked about birds the moment he woke up to the moment he went to bed in the evening. He taught everyone in his family about bird life, from how birds communicated to how they socialised. He could name any bird he saw from a distance, in English and African languages.

The impressive display of bird specimens in the museum is

a marvel to look at. The room where the display is housed is called the Austin Roberts Room, with no mention of Sithole and his work in the field of ornithology alongside Roberts. The history books may not be kind to Sithole by omitting his contribution to the field of science and his extensive knowledge of birds and fossils, but for those who worked alongside him, he left an indelible mark and contributed immensely to this country's scientific history.

Although battered by colonialism and apartheid his whole life, Saul Sithole lived a full life and died at the ripe old age of 89 on 16 December 1997. The museum allowed its staff members to attend the funeral, but his family remembers no white colleagues attending, a demographic he had served so well and worked all his life to please. The love of his life Sophia followed him seven years later in 2004 at the age of 92. Sophia and Saul had been married for more than six decades. At the time of their passing they had two granddaughters and four great-grandchildren.

Afterword

Saul Sithole (1908–1997) was a wonderful person whom I was privileged to know, having met him first when I was almost 20 after leaving school, on the point of entering university for a degree in zoology. At that time, I served as a laboratory assistant in the Department of Palaeontology (the study of fossils) at the natural history museum in Pretoria, then called the Transvaal Museum (now the Ditsong National Museum of Natural History).

Saul commanded respect as a highly skilled member of staff involved with the cleaning of animal skins and bones for purposes of scientific research and for educational exhibitions. In fact, he had begun to do this when *he* was 20 years old. So, there were two things we had in common: we had both begun our careers at the Transvaal Museum, at the same age.

Within the 62 years of Saul's remarkable service at the Transvaal Museum, between 1928 and 1990, he worked closely with Dr Austin Roberts who was both an ornithologist (studying birds) and a mammologist. Roberts trained Saul to clean skins of mammals and birds, and as such he was called a "preparator". In fact, Saul was so good at this that he became the supervisor of all the preparators who were members of the famous Vernay-Lang expedition to the Kalahari in 1930.

As far as birds were concerned, Saul and his colleagues prepared the skins of anything from small common sparrows to large awesome eagles. And as far as mammals were concerned, he prepared bones and skins of animals ranging in size from mice to elephants. The specimens, which he treated with extreme care, went not only into the collections of the Transvaal Museum in Pretoria, but also into those of the British Museum (Natural History) in London (better known as the Natural History Museum), the American Museum of Natural History in New York, as well as the famous Field Museum in Chicago. Those collections of birds and mammals are exceptionally valuable items reflecting the biodiversity of African fauna. They are studied to this day by scientists from all over the world. In fact, they are more precious than gold, at a time when we are facing global extinctions.

There is one thing that I want to say about Saul in the context of Charles Darwin, the eminent naturalist who developed ideas about evolution more than 150 years ago. Saul was doing exactly the kind of thing that Darwin was involved with during the famous voyage of the *Beagle*, a ship which travelled around the world between 1831 and 1836 (stopping briefly in Cape Town). During that voyage of exploration, Darwin collected many birds such as finches from the Galapagos Islands in the Pacific Ocean. Like Saul, he had to skin the specimens. When he was almost 20, Darwin had learned this skill from a "negro", a freed slave called John, who adopted the surname "Edmonstone" and who found a living by skinning birds at the museum in Edinburgh where Darwin was based initially, before going to Cambridge University. Both John and Saul are unsung heroes whose service to science is only now being recognised.

Saul Sithole's involvement with the Transvaal Museum extended also into palaeontology. Remarkably in 1936, at the age of 28, Saul had worked as a field assistant during exciting explorations at the Sterkfontein Caves, with the legendary Dr Robert Broom. At the Ditsong Natural History Museum there is a very valuable photograph of the two of them together with other members of a small pioneering team, just a day after the discovery of a fossil skull (catalogued as TM 1511) which Broom initially called *Plesianthropus* ("almost human"), 2.5 million years old, representing a distant ancestor of all humankind. The historic photograph was taken by Herbert Lang, the very person with whom Saul had travelled on the famous Vernay-Lang expedition into the Kalahari just six years earlier. Without doubt, both Broom and Lang had high regard for Saul, recognising him as part of the team. The photograph is testimony of that. Broom dressed himself in a suit, and Saul wore a tie for that triumphant occasion.

TM 1511 was later called *Australopithecus africanus*, the species which had been initially described in 1925 by Professor Raymond Dart of the University of the Witwatersrand. His naming of this species was based on the discovery of a small skull called the "Taung Child", from the site of Taung in the North-West province. This is the area where Dr Mirriam Tawane grew up. Mirriam obtained a PhD in palaeo-anthropology from the University of the Witwatersrand in 2012 at a time when I was the director of the Institute for Human Evolution there. She went on to join the Ditsong National Museum of Natural History, to fill my former position as curator of Pleistocene fossil collections from the Cradle of Humankind World Heritage Site. It is unfortunate that Saul Sithole was never given the opportunity to fill a similar position of responsibility, recognising that he had learned so much through experience in various fields, including ornithology and mammalogy with Austin Roberts, as well as palaeontology with Robert Broom in 1936.

In 1947, Robert Broom discovered an additional important skull of *Australopithecus africanus* at Sterkfontein. It was nicknamed "Mrs Ples" and classified as STS 5. Saul was working again with Austin Roberts at that time, but would have shared in the excitement of the discovery which was made not only by Dr Robert Broom and Dr John Robinson, but also by Saul's colleague Daniel Mosehle, another unsung hero.

I was appointed Head of the Department of Palaeontology at the Transvaal Museum in 1990 and had the pleasure of presenting Saul with a replica of "Mrs Ples" soon thereafter. He is known to have treasured it in his home in eventual retirement. I told him this: "I have a dream. My dream is to have one replica of Mrs Ples in every school in South Africa, so that learners can see what our distant ancestor really looked like. Mrs Ples was your ancestor, as well as my ancestor! We can be proud of our South African heritage. And we can be proud of you, Saul Sithole, for what you have done at the Transvaal Museum where important collections are curated – including Mrs Ples!" Saul smiled, and I smiled back. We shared a mutual respect.

When Saul was 80 years old, he was appointed a distinguished associate member of the Transvaal Museum. Dr Bob Brain, then the director of the museum, praised Saul for his service within a period of 60 years during which he had been involved not only with the research collections, but also with all of the major exhibitions, including the permanent Austin Roberts Bird Room, and exhibitions in the Genesis Halls which reflect the evolution of the diversity of life within a period of 3.5 billion years.

Saul Sithole has been referred to as an "unsung hero", but in this wonderful book about his life, Lorato Trok is now singing his praises, deservedly so. This is an inspirational book which will stimulate young budding scientists to hone their interests and pursue a career in science passionately, whether they focus on birds, mammals, fossils or other aspects of the diversity of life on our precious planet, earth. Charles Darwin among others would be proud of Saul. To my mind, that is no exaggeration.

Professor Francis Thackeray

Samson (centre), one of Saul's colleagues and other black museum workers outside the museum

About the Author

Lorato has more than 10 years' experience in publishing, writing and story development in children's literature. Lorato was Project Coordinator at the Centre for the Book in Cape Town and also worked as Publishing Programme Manager at Room to Read South Africa. She was the South Africa, Lesotho and Zambia Country Co-ordinator for the African Storybook Initiative, a digital publishing platform for children's stories across Africa. She has written a number of stories in English and Setswana. She also holds qualifications in Languages and Literature (Majoring in Creative Writing) and Advanced Editing. She has presented papers locally and internationally on children's literature and is a creative writing facilitator. This is her second biography for young people and forms part of the series about unsung South African heroes and heroines.

Acknowledgements

I would like to thank the Saul Sithole family for sharing his story and photographs with me. Many thanks also to Nancy Jacobs for introducing me to Saul Sithole and for giving me permission to use her book as a framework to write this story for younger adults.

I would also like to thank Professor Thackery for writing the afterword and for providing the photo of Saul on page 32.

This book would not have been possible without funding from the National Heritage Council. I would also like to thank Biblionef and Jacana Media for their support throughout the process.

Lastly, I would like to thank the Ditsong Museum of Natural History for providing the photos on pages 22 and 28.

Glossary

amadodana Ase Wesile – a musical group synonymous with gospel music in the Methodist church. The group is entirely made up of men.

anthropology – the scientific study of humans, human behaviour, and societies in the past and present

apartheid – a system of oppression against black people in South Africa, classified by the United Nations as a crime against humanity

colonialism – the policy or practice of acquiring full or partial political control over another country, occupying it with settlers and exploiting it economically

Dompas – a document that came into existence with the Natives Act of 1952, or the "Pass Laws". The Act made it compulsory for black South Africans over the age of 16 to carry a "pass book" whenever they were in "white areas".

forced removals – the moving of people from their homes against their will. South Africa experienced a long history of forced removals as a result of racist legislation.

Group Areas Act – legislation enacted by the apartheid government limiting property rights of Indians, coloureds and Africans.

herpetologist – a scientist who studies reptiles and amphibians

indigenous knowledge – unique traditional knowledge systems embedded in the cultural traditions of local communities or distinct societies

Lady Selborne – an area which was situated in the suburb now called Suiderberg, ten kilometres northwest of the Pretoria city centre from 1905 to 1906. It was established in 1905 as a township where Africans could own land before it was identified for white resettlement by the apartheid government.

Mamelodi – a township east of Pretoria

mammologist – a scientist who studies mammals

Mrs Ples – the popular nickname for the most complete skull of an *Australopithecus africanus* ever found in South Africa. It was discovered at the Sterkfontein caves on 18 April 1947.

natives – a derogatory term used by the apartheid government to describe black people

ornithology – the scientific study of birds

paleontology – the scientific study of fossils to determine organisms' evolution and interactions with each other and their environments

Sterkfontein caves – a set of limestone caves of special interest to paleo-anthropologists located in Gauteng, 40 kilometres northwest of Johannesburg. The caves were declared a World Heritage Site in 2000.

stokvel – an invitation-only group of people serving as rotating credit union or saving scheme in South Africa; members contribute fixed sums of money to a central fund on a weekly, fortnightly or monthly basis.

Taung Child – the fossilised skull of a young *Australopithecus africanus*. It was discovered in 1924 by quarrymen in Taung, South Africa.

taxidermy – the preserving of an animal's body via mounting or stuffing for the purpose of display or study.

Sources

Jacobs, N.J. 2016. *Birders of Africa: History of a Network.* Yale University: Yale University Press

Sithole/Zitha Family Archive

Ditsong Museum of Natural History Archive